Looking at the United States

Harcourt
SCHOOL PUBLISHERS

Visit *The Learning Site!* www.harcourtschool.com

Where Is the United States?

READ TO FIND OUT **Where on Earth is the United States located?**

How can you describe where on Earth the United States is? You can tell in which half of Earth it lies. The United States is in the northern half.

You can tell which **continent** the United States is on. The seven continents are the largest land areas on Earth. The United States is on the continent of North America.

This map shows where the United States is on the continent of North America.

THE UNITED STATES

ALASKA

HAWAII

You can tell what other countries the United States is near. Most of it lies between Canada and Mexico. It is south of Canada. It is north of Mexico.

You can use bodies of water to describe where the United States is. The United States lies between two oceans—the Pacific Ocean and the Atlantic Ocean. These two oceans form the eastern and western **borders**, or boundaries, of the United States.

READING CHECK ᵇ **MAIN IDEA AND DETAILS How can you describe the location of the United States?**

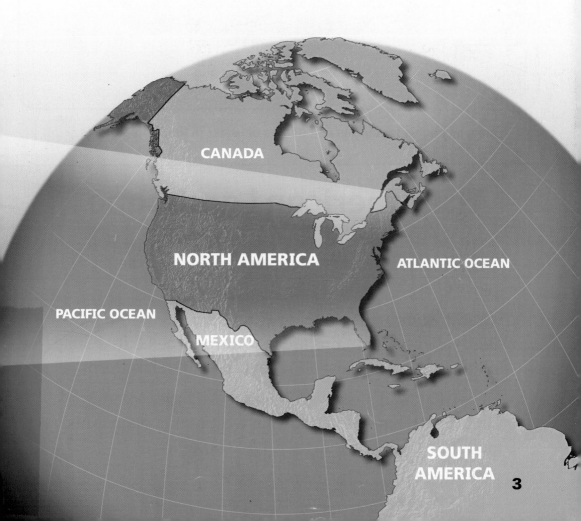

CANADA

NORTH AMERICA

ATLANTIC OCEAN

PACIFIC OCEAN

MEXICO

SOUTH AMERICA

This map shows landforms and bodies of water in the United States.

The American Landscape

READ TO FIND OUT What are the major landforms in the United States?

Landforms are features of Earth's surface. Mountains, hills, valleys, and plains are landforms. The United States is so big that it has just about every kind of landform.

Low, flat land stretches along the Atlantic Ocean. The Appalachian Mountains lie west of this plain. This mountain chain is more than 1,500 miles long.

Hundreds of rivers flow through the United States. It also has many large lakes.

Large areas of low, grassy plains form the middle of the United States. Five huge lakes lie north of this low area.

West of the plains are the Rocky Mountains. They are steep, rugged, and tall. In the West, you can also see deep, narrow valleys called canyons.

READING CHECK **SUMMARIZE** **What kinds of landforms can you see in the United States?**

The Grand Canyon, in Arizona, is one of the deepest and largest canyons in the world.

Climate in the United States

READ TO FIND OUT Why are there big differences in climate across the United States?

How would you describe the climate where you live? **Climate** is the kind of weather a place has over time.

Climate and weather affect the clothes people wear and the jobs they have. They also affect where people live and what they do for fun. There are many reasons for the different climates in the United States.

Weather and climate play an important role in daily life.

One reason is how far a place is to the north or to the south. Places close to the equator are warmer than places far from it. Height above sea level also affects climate. It gets cooler as you go higher.

Oceans and landforms affect climate, too. In winter, oceans warm the land near the coast. They cool it in summer. Mountain ranges block moist ocean air. Places inland from the mountains get less rain.

READING CHECK **CAUSE AND EFFECT** **What are four things that affect climate?**

Natural Resources

READ TO FIND OUT **What are some important natural resources in the United States?**

What is a **natural resource**? It is something found in nature that people can use. Soil, trees, and water are natural resources.

Our nation's rich soil helps farmers grow large amounts of food. Ranchers also depend on soil. They need grasses and crops to feed cattle and other farm animals.

Farmers use soil to grow corn.

An oil pump brings oil to the surface from underground.

People use trees to make homes and other goods. Trees are a resource that can be replaced. People can plant new trees to take the place of the ones they use.

Water is perhaps our most important natural resource. People use fresh water in many ways—drinking, washing, and watering crops.

Some natural resources are deep under the ground. Minerals, oil, and coal are underground resources. They cannot be replaced. It is important to use them carefully.

READING CHECK ŏ **MAIN IDEA AND DETAILS What are some important natural resources?**

The American People

READ TO FIND OUT **Why do Americans have different ways of life, and what brings them together?**

Americans have many different ways of life. Yet many things bring us together. We are proud of our nation. We celebrate holidays such as the Fourth of July.

The first Americans were the people we call Native Americans. In the early 1500s, explorers from Europe began to come to America. Soon settlers came, too.

Large numbers of people moved here to live. They came from all over the world. Almost 300 million people now live in the United States.

These children are proud to be Americans.

10

Until the late 1800s, most Americans lived in the countryside. By 1920, more Americans were living in cities. As cities grew, some people moved to towns and small cities near a large city. Today, more than four-fifths of Americans live in cities and towns.

READING CHECK ⊙ **MAIN IDEA AND DETAILS What are two things that Americans share?**

This map shows the fastest-growing states in the United States.

The Fastest-Growing States

United States Government

READ TO FIND OUT **How is the United States government set up?**

In 1787, the United States was a new nation. It needed a strong form of **government** to decide what was best for the nation's people.

Our country's leaders drew up a plan for the new government. This plan contains our nation's highest laws. It tells how the government is organized. It also describes the rights and responsibilities of citizens.

Our nation's founders worked together to come up with a plan of government.

Our government has three branches, or parts. One branch, called Congress, makes the laws for the whole country. Congress has two parts—the Senate and the House of Representatives. Citizens vote for the members of this branch.

Another branch carries out the laws. The leader of this branch is the President of the United States. Citizens vote for the President.

The Supreme Court and other courts make up the third branch. These courts make sure that laws are fair. Many judges are chosen to serve in their jobs.

READING CHECK ☽ **MAIN IDEA AND DETAILS How is the United States government set up?**

United States Economy

READ TO FIND OUT **How does the United States economy work?**

The **economy** is all the ways people use resources to meet their needs. The United States has a free market economy. People can make, sell, and buy almost anything they want.

Natural, capital, and human resources are all part of the economy. Capital resources are money, buildings, and machines. Human resources are workers.

This customer and this worker are both part of the economy.

Many countries have their own money.

People are also part of the economy. People become part of the economy when they buy and sell things. The price people pay for something depends on two things—people's wants and how much is for sale.

Farming was once the main job. As cities grew, making goods became more important. Today, most people have service jobs, such as teachers and nurses. They are paid to do things for other people.

READING CHECK ☼ **MAIN IDEA AND DETAILS How does the United States economy work?**

Understanding Regions

READ TO FIND OUT **Why do people divide places into regions?**

A **region** is a place with something that makes it different from other places. Dividing places into regions makes understanding the world easier.

Each place in the United States is part of many kinds of regions. A region may be large or small. North America is a region. So is your neighborhood.

This map shows features of many regions in the United States. The route in red passes through each region.

Chinatown in San Francisco, California, is a cultural region.

Regions can be divided into four major types. A place where people share a government is a political region. Physical regions are based on the landforms found there. An economic region is based on the way people use resources. A cultural region is a place where people share certain ways of life. All regions develop and change over time.

READING CHECK DRAW CONCLUSIONS **Why do people divide places into regions?**

This map shows the five regions into which the United States is divided.

United States Regions

READ TO FIND OUT What kinds of political regions does the United States have?

The United States has many kinds of regions. The states are often grouped into five large regions. The states in each region are similar in many ways.

Every state is a political region with borders and a state government. Each state is divided into many smaller political regions called **counties**. Each city and town in a state is also a political region.

The United States has three levels of government. Each level collects taxes to pay for services to citizens.

The national government gives services to the whole country. Delivering mail is one of these services. State governments give services to the people who live in their state. Taking care of state roads is one of these services.

Counties, cities, and towns give services to the people who live in them. They give police and fire protection.

READING CHECK **SUMMARIZE** **What kinds of political regions does the United States have?**

Local taxes pay for firefighters and their equipment.

Neighboring Countries

READ TO FIND OUT With what other regions does the United States share North America?

The United States shares North America with several neighbors. Canada lies to the north. It is our largest neighbor. In fact, Canada is the second-largest nation in the world. It has big cities, mountains, and forests.

The north of Canada is very cold. Most people live in the south, within 100 miles of the United States border.

Like the United States, Canada has many different landforms.

Mexico City is the capital of Mexico.

Mexico is our neighbor to the south. Mexico is made up of states. Like Canada, it has large cities and many different landforms.

Central America is south of Mexico. It is a narrow strip of land. This land connects North America and South America. This region is divided into eight countries.

Islands in the sea are our neighbors, too. Greenland is off the coast of Canada. It is the world's biggest island. Many other islands are south of the United States and east of Mexico.

READING CHECK ŏ **MAIN IDEA AND DETAILS** **With what other regions do we share North America?**

Activity 1

Choose the term that correctly matches each definition.

a. region   d. economy   g. government

b. border   e. landform   h. natural
                                            resource
c. climate   f. continent
                                        i. county

___ 1. a feature on Earth's surface

___ 2. all the ways people use resources to meet
       their needs

___ 3. something found in nature that people can use

___ 4. a line that shows the end of a place

___ 5. the kind of weather a place has over time

___ 6. one of the seven largest land areas on Earth

___ 7. an area with at least one feature that makes it
       different from other areas

___ 8. a system for deciding what is best for the people
       of a nation

___ 9. a smaller political region in each state

Activity 2

Look at the list of vocabulary words. Categorize the
vocabulary words in a chart like the one on the next page.
Then use a dictionary to learn the definitions of the words
you do not know.

Activity 2 (continued)

hemisphere   judicial branch   municipal   patriotism

climate   gulf   treaty   profit

government   mineral   province   mouth

equator   checks and balances   territory   immigrate

humidity   commonwealth   supply

constitution   landform   tropics   basin

prime meridian   nonrenewable   rain forest   population

natural resource   majority rule   tundra   demand

democracy   sea level   environment   canyon

continent   republic   economy   rural

renewable   interdependence   plateau   manufacturing

legislative branch   natural vegetation   conservation   temperature

free market   urban

relative location   ethnic group   source   service industries

groundwater   modify   culture

executive branch   technology   factors of production   precipitation

communication   suburb

border   county   tributary   region

industry   county seat

		I Know	Sounds Familiar	Don't Know
○	plateau			✓
	gulf		✓	
	democracy	✓		

(⭐Focus Skill) **Main Idea and Details** What are some of the things that bring Americans together?

Vocabulary

1. What type of area is a **county**?

Recall

2. Why do Americans have many different backgrounds and ways of life?
3. What are the jobs of the three branches of the United States government?
4. Why do governments collect taxes?

Critical Thinking

5. In how many different types of political regions do you live? Come up with at least four.

Activity

Draw a Map Draw a map of your classroom. Divide the room into regions based on what is done in each area. Then explain why you divided the room the way you did.

Photo credits Front Cover Photo Researchers, Inc; 5 Getty Images; 6 (l) Stephen Wilkes/Getty Images; (r) Getty Images; 7 (l) Jonathan Nourok/ PhotoEdit; (r) AP Images; 8 Ric Ergenbright; 9 Inc., DesignsPics/ Index Stock; 10 Creatas/Age Fotostock; 12 The Granger Collection, New York; 14 Gayle Harper/ In-Sight Photography; 15 Getty Images; 17 David R. Frazier Photolibrary/ Alamy Images; 19 Mark Richards/ PhotoEdit; 20 K. Yamashita/ Panstock/ Panoramic Images/ NGSimages.com; 21 Randy Faris/Corbis